FLAX OIL AS A TRUE AID AGAINST ARTHRITIS, HEART INFARCTION, CANCER AND OTHER DISEASES

(FETTE ALS WAHRE HILFE GEGEN ARTERIOUSKLEROSE HERZINFRAKT KREBS)

Dr. Johanna Budwig

Canadian Cataloguing in Publication Data

Budwig, Johanna, 1906-
    Flax oil as a true aid against arthritis, heart infarction, cancer and other diseases

3rd ed.
Translation of: Fette als wahre Hilfe gegen Arteriosklerose, Herzinfarkt, Krebs.
ISBN 0-9695272-1-7

    1. Linseed oil--Therapeutic use. 2. Diet therapy. I. Title.
QP752.F3B82 2000   615.8'54   C99-901626-1

APPLE PUBLISHING CO LTD
5275 Clarendon St
Vancouver BC
Canada
V5R 3J7   Tel: 604/438-6929   Fax: 604/438-6939

E-mail: books@applepublishing.com
Website: www.applepublishing.com   PRINTED IN HONG KONG

FOREWORD

Upon returning from Easter holidays, the proof for this third edition was waiting for me on my table. At the same time, I received an important professional publication—the May 1994 issue of the *Townsend Letter for Doctors*. In this issue, my comprehensive, groundbreaking report *"Natural Occurrence in Fat Systems in Humans"* was published. This report documents my new findings about the oxygen consumption of the human body.

In the same issue, it also noted that scientists who confirm my new findings are under scrutiny by industry interests. Both writers Martin J. Walker and Jule Klotter comment on how these pressures, common in Europe, are also happening in North America. They write about the present state of orthodox medicine as 'fighting over lost ground.' We question the goal of the members of orthodox medicine—is it for our benefit?

The implications of my 'lectures for laymen' are only fully understood when we can see the whole picture: huge differences of opinion exist between scientists as how we truly help people worldwide. The time for truth is now.

Amazing response from around the world to these articles shows the importance of this work. As a rule, a breakthrough like this takes years before it is accepted by members of the 'old school.'

I would like to thank the publisher Mr. Alexander Pazitch for his commitment and patience in putting together this third edition.

Flax Oil as a True Aid

for Arthritis, Heart Infarction, Cancer & Other Diseases*

LADIES AND GENTLEMEN

I want to say how much I appreciate your coming here in such
great numbers. But firstly I must thank and acknowledge the
organizers here, the *Verein für Volksgesundheit* (the Association
for Public Health), who during their long years of pioneering,
often under considerable harassment concerning the clarification of the fats question, have remained steadfast. They have kept
the faith despite being badgered, when the question of the truth
arose, particularly concerning the question of fats. I also wish to
thank the many friends here in this hall, who during the last five
years have been staunch supporters and seekers of truth, even
when they were unwillingly placed in the very centre of
controversy. However, I am no less grateful to those of you who
are here for the very first time to learn something about this
question which has assumed enormous relevance at present—
this simple question concerning food fats. Why have fats become
so very significant today?

Anatomically speaking, heart infarction observation studies
reveal nothing abnormal in the picture except for solid fat, which
encircles the normally lean heart muscle, confining and disturb-
ing the heart's action.

In rheumatic disease, the only factor which distinguishes the
diseased muscles from the healthy ones is isolated fat—a clear

---

* Lecture was held on November 2, 1959 in the Zürich Conference
Centre and in similar form in other places in Switzerland.

"No!" to this substance. At the international *Ernährungskongress* (Congress of Nutrition) in Paris in 1957, 900 expert delegates from countries worldwide heard that the latest research results, in the Cancer Research Institute in Paris—the only research institute equipped with the largest state-of-the-art electron microscope in existence—showed that the only substance which characterised the cancerous cells, as opposed to healthy ones, was isolated fat; the formation of fat in the cell nucleus, cell body and cytoplasm. It was the single distinguishing feature of cancerous cells in contrast to healthy ones. Of course, when interpreting this, nobody is yet prepared to acknowledge its corresponding significance. Instead, the following course of action is proposed:

"We should now investigate reptiles and other lower life-forms for similar isolated fat in the nucleus and cell, to ascertain what degree of physical degeneration man has reached. . . ."

In connection with this I had the opportunity at the Paris Congress to ask, "Why not investigate the phenomenon itself?" Isolated fat, a clear "No!" to this substance. The President of the Noble Prize Committee in Stockholm had already wondered whether the whole cancer problem could be solved by investigating 'lipolytic substance'. 'Lipos' means 'fat'. The lipolytic substance causes fat to dissolve again. When living tissue rejects some fats, the body isolates them—and this is the crux—and deposits them in places where fats are not normally found. Bauer, whose book *"Das Krebsproblem"* (The Problem of Cancer) has made him world famous, wrote in his 1966 edition "everything points to fats playing an enormous role in this problem." But the statements "Fat can dissolve tumours" and "Fat can cause tumours" are so contradictory that it is not at present possible to draw any viable conclusions which can be of benefit to cancer research. It is therefore not beyond the bounds of credibility that the appropriate division in the *Bundesinstitut für Fettforschung* (Federal Institute for Fats Research), was moved to begin examining the question more closely. And initially we confirmed that proofs for fats were lacking.

In 1949–50, it was not possible to distinguish between the fatty acids found in olive, linseed and sunflower oils. The specialists in fats with us today in this hall will confirm this. The *"Kennzahlen Methoden"** gave no surety against misleading results. No direct reactions were known, such as those which have existed for decades in the protein field or in testing the various sugars. It is not the first time in the history of chemistry and medicine that the presentation of clear, new, beneficial, usable and extremely sensitive, conclusive evidence reactions has opened up new territory for research in general. Through the development of paper chromatography in the field of fats, that is, by means of new tests of various fatty substances on paper, which I first carried out in 1949, without knowing how great their effect would be on the world of medicine, it was possible to analyse a milligram of a blood to the thousandth degree. That meant that the fat in a drop of blood could now be broken down into its various constituents and precisely catalogued by its diverse fatty acid components. And, during these studies, several publications appeared, connected with protein work really, asserting the following:

The absolute lack of any conclusive detection methods or reactions for determining abnormalities in the metabolism of fats in living humans is very noticeable in medicine generally. We can, regrettably, only study

---

* Identification Number Methods

abnormalities of fat metabolism in the final phase, i.e. upon the patient's death. [1949]

Examination of fats taken from blood then began and inaugurated a tremendous new field of productive research which is being intensively prosecuted in America, England, Belgium, Holland and all over the world. A surprisingly high number of correlations between fatal diseases and the metabolism of fats have been discovered. To solve this, it was only necessary to perceive that 'fat' is not always 'fat'. The whole situation was instructively summed up in a comment made by a French professor at the *Ernährungskongress* (Congress of Nutrition) in Paris. During the congress he said, "What's all this about fats? Nobody wanted to know a thing about them before —fat was an oily, greasy substance that no one knew what to do with. And suddenly everybody's talking about it now, whether they're in the groups handling carbohydrates, protein or other metabolic factors. Everybody's on about it!" The Paris press reported: *"La margarine sur la sellette!"* – "Margarine stands indicted!" Two weeks earlier in London, the Daily Express had said: "The fat in your frying pan can be deadly!" A conversation was reported between a husband and wife; HUSBAND "Give me that steak, but throw out the fat." WIFE "But that fat cost me 36¢!" HUSBAND "Is not your husband worth 36¢ to you?"

In this way, it became apparent that throwing away harmful fat, rather than simply eating it to save money, was a far better idea. This is the crux of the matter. Are all fats the same? Why had fat become so pressingly important at that time? How is it that fats can both cause tumours to form and also to dissolve? How is it that fats both cause and present heart infarction? What is new in this field? The answer to this, and exact research into what fat actually is and actually does, has only been possible since 1950 and it is urgently needed, especially today in the era of, on the one hand the industrial solidification of fats and, on the other, the enormous importance of natural, electron-rich, vital, highly unsaturated fats.

With your permission and for clarification purposes, I'm going to say a few worlds of explanatory introduction about what fat actually is. Fat consists of glycerine. Glycerine has three 'arms'. A chain of fatty acids is attached to each arm. This chain of fatty acids has, with butter for example, 4 links to it; that of coconut butter and palm nut oil, 14 or 16 links. These naturally occurring saturated fats cannot be turned into energy by the body without the presence of essential fatty acids, but, at the moment, it is those in a different group which play a larger role, particularly when there is a basic lack of the essential, highly unsaturated fats. What's going on? I'll come to that later!

What is unsaturated fat? Upon analysing fatty substances, we come to the fatty acid chains with 18 links, and can observe that the links are, in places, not so firmly attached to each other. The chain is fragile there, loose; it absorbs water easily—as if you were to fray a smooth silk thread in one place and then draw it through water. The frayed part absorbs water, or dye, more easily. In the same way, these fatty acid chains with their weak, unsaturated connections, form protein associations very easily. The fatty acids become water soluble through this association with protein. There are unsaturated fatty acids in, for example olive oil, which has itself only one unsaturated connection. This fat is not harmful, but neither is it the most beneficial which we

need in Europe. The Russians in the Ukraine, whose diet is
different, fall into another category. What we need in Europe
today, in Germany as well as in Switzerland, America and
France, what we really need, are highly unsaturated fats. The
moment two unsaturated double links occur together in a fatty
acid chain, the effects are multiplied and in the highly unsaturated fats, the so-called 'linoleic' acids, there is generated a field
of electrons, a veritable electrical charge which can be quickly
conducted off into the body, thus causing a recharging of the
living substance—especially of the brain and nerves. To stabilize
these fats the unsaturated connections, which require oxygen so
actively, are destroyed by industrial handling methods. But it is
exactly those highly unsaturated fatty acids which play a decisive
role in the respiratory functioning of the body. Without these
fatty acids, the enzymes in the breath cannot function and we
asphyxiate, even when given extra oxygen, as for example in
hospitals. The lack of these highly unsaturated fatty acids
paralyses many vital functions. Primarily, it cuts off the air we
breath. We cannot survive without air and food nor can we
survive without these fatty acids—that was proven long ago.
Now, and I address these remarks to the directors present today
from the fats solidification industry, I regret to say that no one
has known any better since 1902, than to start treating fats so as
to make them more easily handled, commercially marketable,
longer lasting and slower to turn rancid, also to make them more
easily spread, because the people in our climatic zones apparently
want it that way. These unsaturated fats have been chemically
treated so that their unsaturated qualities are destroyed, the field
of electrons removed. Their ability to associate with protein and
thereby to easily achieve water solubility in the fluids of the

living body—all this is destroyed. These fats are no longer active at the surface and capillary level, that means they can no longer flow into the capillaries. Medically speaking, one says "the blood needs thinning". The solid fats which are not water soluble and cannot associate with protein are no longer capable of circulating through the fine capillary networks. The blood thickens and circulation problems arise. I want to say what this is, before going on to individual organic abnormalities and to the enormous complexity of the effects of the fat metabolism on all the vital functions. Solidified fats behave differently according to which oils are taken. When natural oils such as rapeseed or olive are used, starting substances which are rare today, then the solidified fatty acid chain of 18 links is similar to that of pork fat, and that would be the best possible scenario. When peanut oil is taken, using solidified peanut butter (you may have heard that such an item is available commercially), this is far worse, as the fatty acids in peanuts consists of 20 links. The situation is even worse for the body when one considers fish oil. The chains of fatty acids in fish oil have 26, 28 or 30 links, with numerous unsaturated connections. In this form, fish or liver oil is ideal for the body due to the numerous unsaturated connections of the electron groups and their ability to associate with protein and bring about the formation of new material. This is due to the electrical dipolarity between fat and protein during cell growth and is indispensable for the regeneration of any body substance, in adults too. Fish oils which have been rendered oxygen-inert by means of solidification and which have had their electron fields destroyed are no longer able to carry out the above listed reactions as part of the fats metabolism within the living organism. The correlations between these things were not recognized and the fact that such methods of preservation have been used by the fats industries is, however, not a reason to hurl allegations at these industries. But the fact that scientists who are centrally placed regarding this research and who are aware of these connections, that they then believe that they can do away with this truth by attempting to suppress the continuation of this scientific work, this is indeed a case for recrimination. However, my feeling of satisfaction in this connection is far stronger than the latter—a feeling of satisfaction that whatever else, progress in the scientific direction just mentioned has been clearly illustrated during the last nine years. I am satisfied that this is a truth of great importance to mankind and a truth in defense of which there are some people who are willing to make sacrifices and stand up and be counted. This truth, however, is being suppressed by the power of money and other strongly placed means of leverage. But this truth will prevail and can no longer be denied. It can be summed up very simply in two points.

The metabolism of fat has such an extensive effect on the vital functions and every individual organ—quite simply on life itself, including the generation of new life, that the lack of unsaturated fats is no longer acceptable. Well-meaning methods of treating fats to give them a longer shelf-life, but which destroy their fatty substances, must be changed. The metabolism of fat affects each and every organ. Any patient with liver and gall bladder problems is fully aware of how fat makes him feel. Medical treatment prescribes the eating of less amounts of fat because it has been observed that the patient cannot digest it. However, if he is given beneficial fats within the definitions I've just summarized, that is, highly unsaturated fats, then he has no digestive trouble at all. It is best to use threefold unsaturated fats prepared from flax seed oil, together with the substance which easily dissolves them, and that is cottage cheese. Various highly trained and educated individuals are dismayed and irritated by the fact that serious medical conditions can be cured by cottage cheese and flax seed oil. The diseased organ is in no way harmed by these substances. There is only one other important point to be observed. The usual methods of preserving foods are based on the addition of 'oxydation'—inhibiting substances which render inert that process itself, the combustible process in the foods. The majority of preservatives can be labelled as respiratory poisons in that their effect is to block the combustion of fat—its continued interaction with oxygen. When, in such a situation as described, we supply respiratory activating fats and prevent the ingestion of preservatives, which are respiratory poisons, then great numbers of patients who have been given up as hopeless cases by many clinics, will recover their health. Some days ago, while lecturing in Meilen, I was pleased when afterwards, a member of the audience stood up—someone I had neither encouraged nor even seen or spoken to previously—and whose husband had a tumour of the lung and whose daughter had a hopeless case of psoriasis and degeneration of the joints' cartilage substance, and had been pronounced as incurable by several clinics. There was nothing to be done except bed care and nursing. The family's son also had several minor health problems. This woman, a wife and mother, stood up and declared that everyone had regained their health after having changed their nutrition to include the oils and proteins I recommend. It is rather odd for you to have to believe this, and it is very difficult for me to have to encompass the entire complex system which is affected by the metabolism of fat, and which I call the 'fats syndrome', and its real connections with the

metabolism of fat—and all that within 20 minutes! But I want
to try to explain it in telegram style. Those of you who wish to
know more will find corresponding literature available.

In the formation of mucous secretions, it is the particular
interplay of the unsaturated, surface-active fats, those fats which
penetrate to the surface, and of protein, which is paramount.
There is no gland, neither the liver, the pancreas nor the glands
in the upper body cavities, the tonsils, salivary glands and others,
nor the lower body glands, which can produce mucous, when
a lack of unsaturated fats is present. The drying out of mucous
membranes is nowadays widespread and often a course of
complaint. This problem is easily remedied by reintroducing the
body to natural, highly unsaturated fats, and this is significant. The
fats which are alien to the body block the metabolism of other
fat in the delicate glands, capillaries and filter stations. Such
solidified, heat-treated oils must be avoided. Only then will the
vital functions return to normal—the glandular secretions within
a very few days, sometimes 24 hours. The basic problem with
diabetics is really an impairment of the fats metabolism system,
and not that of the conversion of sugar. The sugar assimilation
problem is only secondary.

A lack of highly unsaturated fats is particularly noticeable in
connection with brain and nerve functioning. An adjustment in
diet to one with oil and protein contents high in unsaturated fats
brings the best result in children. I have often observed this when
called in to treat cancer patients. In general, I recommend that
the whole family adjust their food intake so that they use
the optimal, natural fats. As for children whose scholastic
performance is often below standard—and it's usually the case

in families where the parents don't eat correctly—the results of an optimal fat intake normally begin to show themselves in school marks being bettered by not only one, but two levels.

The heart's function is affected in three ways when there is any abnormality in the metabolism of fats. The ingested fats are transported via the lymph system. Before the blood, that is the venous blood which the body has variously utilized and which is low in oxygen; before this flows into the right-hand ventricle of the heart, each heart beat deposits lymphatic fluid into the blood, that is body fat, directly from the digestive system. The blood that fills the left hand ventricle comes from the lungs and is freshly oxygenated. The electrical potential differences in the relation of the current between the newly fat-laden venous blood in the right-hand ventricle and the oxygenated blood in the left-hand ventricle are directly involved with the generation of the heart action currents. This is immediately recognizable anatomically, and can be accurately measured. If then, there is a lack of new electrical impulses during this recharging with fat and if the fat is loaded with inert and paralysing fats, then the heart says "No!"—it rejects these fats which build up in the coronary vessels and then on the entire muscle. In addition, as has long been proved scientifically, the heart muscle simultaneously suffers the lack of a substance which plays a leading role in oxygenation, breathing and the regeneration of the heart muscle itself. This substance, I only mention it out of scientific interest, is known as 'cytochrom-oxydase' and corresponds to the substance which Warburg called *"das gelbe Atemferment"*.* Thus, when the heart by means of isolated, solidified fats, shows that the wrong kind of fats are being ingested, then there exists a lack

* The yellow respiration enzyme

of optimal fats essential for the heart action and functioning. And it is exactly in this situation that highly unsaturated fats are also lacking in the blood. Oxygenation of the blood via lungs is hindered and the heart is therefore forced to pump the same amount of blood through the body three or four times before the tissues are properly supplied with oxygen. Now, as a third element in this situation, there appears the fact that only natural fats can easily pulse their way through the finest of the capillary networks. The solidified, inert bulk fats only act as further hindrances in the blood. In America, it has been proved that fasting animals immediately release fats into the larger blood vessels, aorta and arteries, when their previous diet contained only saturated fats. The condition know as "hardening of the arteries" has long been medically understood as an abnormality in the metabolism of fats—as an isolation of fat. When these same animals are fed a corresponding amount of highly unsaturated fat before fasting, then such isolation of fat during fasting does not occur. In the last few years, particularly last year, some interesting tests have been carried out on rats. Yesterday I read in the newspaper *"Die Weltwoche"* that our children are reaching physical maturity ever earlier and this is beginning to give rise to anxiety. Criminality and other social problems have been connected with this factor. The tests on rats and mice showed in experiments that when the animals from one litter are fed partly with solidified fats—you can use solidified peanut butter, you don't have to give them the worst fat available—and others are fed naturally highly unsaturated oils, it is observable that the animals lacking the unsaturated fats and which have been given solidified fats, mature physically much earlier. The mating cycle is disturbed. It was further written "they will mate intensively."

The young animals in this condition will rub their paws raw on the cages; but not those with a normal metabolism. The mating cycle is disturbed, with the young males being impotent and the females giving birth to dead and dying young. Some weeks ago, an important article came into my hands. The origin was Professor Dam in Denmark. Through his work he actually belongs in the margarine union's camp, but after many battles he has decided to honour truth after all. He has published his findings that, with young male animals, their testicles and neighbouring glands are defunct of semen within eight weeks of being fed solidified fat—solidified peanut butter. Abnormalities appear in the skin, hair is lost and the animal is altogether in bad condition. This is also confirmed by the extreme changes in the kidneys. Another test used only two foods as disease-causing agents—items such as those obtainable from candy shops, and sausages—nothing else. The animals showed comparable symptoms of a nutritional deficiency and began to chew on each other as an indication of this. Some years ago, I was attending a cancer patient in Lucerne when the mother presented her two children, 12 and 14, to me. The children were very over fed with the boy being unhealthily obese and the girl having an unhealthy appearance. The mother complained to me about her children's bad behaviour. I felt for her. And then I lay the blame on the food she prepared. The woman denied this. Her daughter wouldn't obey her or behave in a friendly manner to her at all. After a long discussion of other topics, I said my farewells and turned to go. The daughter slipped through a side door and caught me in the hall, inquiring in a whisper: "What's the name of the medicine I can take to make me a good girl? Can't you give me some?" I gave the girl a mixture of flax seed and honey, rich in highly unsaturated fats. That young girl could not have better described the overall situation today—the youth problem. It's not true that the younger generation are not willing.... Altogether I put the main blame on the older generation which very often, for the shameful sake of commercialism, refines food and curtails their content of essential fats. The entire situation is as illustrated by our younger generation today. Not all moral behaviour problems can be solved by means of eating fats and oils—but I assure you that the effects on our society of a normalization of fats in our nutrition are tremendous. Married life is often complicated by difficulties which are clearly connected with incorrect diet and its repercussions on the metabolism of fats and in turn, its influence on the couple's sex-life. And human sexual relationships can be very positively influenced, as experience has often shown, by highly effective natural fats in the diet.

One could, in connection with diseased organs, consider the individual cases in the extensive fields of skin ailments. Kidney disease is very indicative here. Professor Dam, of Denmark, who I've just mentioned, has published that in the case of these young rats, within eight weeks there were extreme abnormalities of function observable and histologically, atrophy of the kidney substance itself. The fine filter systems in the kidneys consist of extremely thin lipoid membranes which contain unsaturated fats existing there because, as filters, their surface activity is essential. The membranes stretch easily and are not cohesive, as is tumour tissue. I don't want to elaborate on each and every organ's function and its relation to disturbances of the metabolism of fats but at the end of these remarks I will discuss the extensive areas of disturbances in growth and proliferate cells.

In growing cells, we find a dipolarity between the electrically positive nucleus and the electrically negative cell membrane with its highly unsaturated fatty acids. When the cell divides, it is the cell nucleus which begins this. The cell body and the daughter cell are then separated and tied off by the lipoid membrane. When a cell divides, its surface area is larger and must, of necessity, contain enough material in this surface with its fatty acids, to be able to divide the new cell completely from the original. Normal growth is always distinguished by a clearly defined course of action. In all our skin and membranes, in that of adults too, there are continual growth processes. The old cells have to be shed with new ones being formed underneath. When this process is interrupted, it means the body is beginning to die.

Cell life is equally dependent on the functioning of the unsaturated fats in the membrane, the external skin of the cell. The entire range of foods, particularly those concerning the metabolism of water and its assimilation, are all dependent on a proper functioning of the intake of highly unsaturated fats in their interplay with the protein of the cell plasma. The protein substance too has to be continually renewed. It has already been clearly established that carcinogens, or chemical substances which are known to cause cancer, attach themselves to the parts of the cell which reproduce the protein substance, and in the external lipoid membrane, which is where highly unsaturated fats are especially localized.

With regard to the preservation of structure in the living body, the dipolarity of the electrical field between fat and protein is of fundamental importance. If this dipolarity between highly unsaturated fats and the sulphur-containing protein substance is

destroyed, for example due to fats having been solidified before
being ingested, that means their electrical charge is removed so
that the counter-polarity is missing for the maintenance of a
voltage field. In short, the battery is empty. It is exactly the same
with a car battery—when one pole is removed, the current no
longer flows. A well-known American researcher has already
tried to free medicine from diagnosing interconnections purely
from localized observation. His name is Selye and his renowned
book is "The Adaption Syndrome". He observes that a healthy
organism is well able to adapt to all manner of alien circum-
stances—to cold or heat, to overstimulation of the nerves or to
any other demands made upon it. He calls these demands 'stress'
and writes that everything depends on being able to recharge the
battery of life. "I have the impression sometimes, that this life-
battery is not being properly recharged." And with regard to this,
the significance of unsaturated fats rich in electrons is especially
clear. It is these electrons which are capable of recharging our
batteries. All the symptoms of a dead battery are always closely
juxtaposed with the manifestation of the condition we call a
tumour, or cancer or excessive growth, that is to say—myoma.
Such tumours are only one manifestation in the whole complex
system just described.

The formation of tumours usually happens as follows: In those
body areas which normally host many growth processes, such as
in the skin and membranes, the glandular organs, for example,
the liver and pancreas or the glands in the stomach and intestinal
tract—it is here that the growth processes are brought to a
standstill. Because the required dipolarity is missing, due to the
lack of unsaturated fat, the course of growth is disturbed—the
surface-active fats are not present, the substance becomes inac-

tive before the maturing and shedding process of the cells ever takes place. I emphasize: It is not correct to regard the problem of tumours simply as a problem of too much growth and thereby to instigate all manner and means of growth inhibiting treatments, such as radiotherapy, hormones and cortisone. I am well aware, that this is a daring statement. I knew this when I said the same thing some years ago on radio in 1956. But these things must be said loud and clear so that those who are suffering can finally get effective help. In substantiation of what I've said, and to possibly facilitate the making of a difficult decision concerning all this, I may add: When the Zentralausschub fur Krebsforschung (Central Committee for Cancer Research) in Germany, represented by three professors, tried to take legal action against me, for just this statement I have made before you, the presiding judge said: "Doctor Budwig's documents and papers are conclusive. There would be a scandal in the scientific world, because the public would certainly support Doctor Budwig." He advised the professors to withdraw their accusations but they were obstinate and did not comply. Then even the University's Vice Chancellor, himself a jurist, became involved. The entire case was declared null and void in order to avoid a public outcry.

If that which I maintain to be true were too far-fetched, it would have turned out very differently. It is the only conclusion which can be drawn from the remarks I have made and I flatly declare that the usual hospital treatment today, in a case of tumorous growth, most certainly leads to worsening of the disease, or a speedier death, and in healthy people, quickly causes cancer. It is for these reasons that, when counselling on nutrition, which I often do with patients, I cannot allow other methods such as growth inhibitors, to be used. The oil and protein intake which

I advise consists of simple food, mainly the most active fat I know—flax seed oil—in easily digestible form accompanied by cottage cheese and, as a consolation for the gourmets among you, in appetizing, tasty meals. It has been tried for many years by the Swiss who find it palatable when properly prepared. There is a cookbook available for those new to it. This simple food revises the stagnated growth processes thereby naturally causing the tumour or tumours present to dissolve and the whole range of symptoms which indicate "a dead battery" are cured. In a short time the patient feels well again. It is preferable, however, not to wait until three or four doctors have pronounced a tumour as incurable—but rather your principle task is to recover your health completely by optimal nutrition.

I am delighted to see that recently in Germany, young people and athletes have become aware of these matters. Reports of athletic achievements often mention that the Russian athletes have analyzed closely their oxygen intake for their perform-ances. It is a clear aspect of a central problem, not only for the hospitals but also for the experts in health reform.

Many viable ideas have been put forward in the matter of healthy living styles. Physical exercise outside in the fresh air is certainly beneficial, and natural foods as I have described them are always the best. But all this counts as nothing when the question of fats is not understood. Today in our present complicated western civilization with its abnormalities and excesses, we need optimal fats to recharge our physical and mental batteries. And I suggest that those of you who doubt whether to take this step or not—risk it for just three days! And if, after those three days of adjusted food intake with its beneficial, optimal natural fats and excluding all other, indigestible, harmful fats and the respiratory poisons found in food preservatives—just consider how you ingest these substances—and if, after following this adjustment for three days, you then return to your previous food intake patterns, which I doubt, then you can write to me. I will refund you all expenses incurred. I have often promised in individual cases: "I'll give you SF100, if you really want to return to your previous foods"—this in particular with very sick patients when it is not possible to hold a conversation or give a lecture, when it's simply a matter of easing someone and helping him to overcome any prejudices against these natural oils. There will be a chance after a short break for you to put your questions, which may relate to this practical aspect. But please think carefully, it doesn't take a lot of time—those saturated fats and the semi-saturated fats are inert and are of no use to your body. It has nothing to do with following a particular strict dietetic regime—it is simply a matter of recognizing what is, at this moment, of optimal value for us all, of recognizing what is food with little benefit and also highly toxic. This decision concerning the substances you eat is entirely up to you. I have often experienced that when, during discussions someone asks a question, there then comes a long list of food items with the enquiry "Is this good for us? Is that or the other acceptable or harmful?" Please do not ask me about the particular products of any one firm or other. Think about fundamentals and buy the oils which are recognizable under their own descriptive names. I'm always very sceptical of imaginative names and you are also more assured of better value for money when you know what you are buying—sunflower, flax seed, sesame or poppy seed oils. I can't give you any information about fats or oils which have fancy names—because that is a field all in itself. If I were a housewife I would want to know what I'm getting for my money—and good fat saves money.

Animals which have been fed with solidified fats or saturated, inert fats, eat six times as much fat and six times as much food. A lot of money is saved by buying the right kind of fat. I don't know whether there's anything to be added on the purely practical side. It is simply a matter of including natural highly unsaturated fats in your food intake. And as our Western countries have no desire to forego easily spreadable fats, neither in cooking nor on bread, the health food shops are also able to offer flax seed oil in an easily spread form—a spreading fat with about 30% flax seed oil. You don't have to buy this fat, but experience has shown that it is more convenient to have available a beneficial, optimal fat in spreadable form.

A few words about flax seed because the risks involved are high. The optimal fats are, of course, the most oxygen-active ones. When flax seed is rough-ground, the beneficent fatty acids, that is the threefold unsaturated ones, the very best ones, will spoil rapidly—within 10 to 15 minutes. And if you rough-ground flax seed from a health food store, by the time you get it home, its goodness has been destroyed and the oxidation waste products are harmful, particularly so, the older they are. Caution is advised with flax seed. There is one product which contains honey as a preservative. And I feel that flax seed products which contain honey as preservatives for the rough-ground flax seed are always the best. Eating whole flax seed is a waste of money because the body cannot extract its goodness. If you keep an eye on your digestive processes, you will observe this clearly for yourself.

Most of the whole seeds cannot be digested and are simply passed through the body and then excreted. Nutritionally speaking, flax seed as found in Linomel, is much more highly recommended. Food should be eaten with enjoyment.

It is easy to put into practice which I have said. You only have to be aware of where hidden, harmful fats can be found—perhaps for example, in cakes and pastries and biscuits, sausages and luncheon meats. Where can respiratory poisons in the form of preserving agents be best avoided? In this connection, I'm directly addressing the housewife—think carefully. When in Dusseldorf, while treating a patient and acting as advisor to the household in question, a new girl started in the kitchen there. She also did the cooking and had previously worked in the catering industry. During my first visit the whole matter of new dietary items appeared to be causing her some difficulties. Upon my return 14 days later, this employee had adjusted wonderfully and had readily grasped the essentials. When I asked how things were going in the kitchen, the girl replied: "Very well indeed, it just takes a bit of extra thought." The oils and proteins I recommend are not a new scheme for any special diet regime. It will be demonstrated how these essential, naturally high-active fats can be re-integrated into our eating habits and how to make tasty meals with those foodstuffs rich in healthy fats. In this way, fats really do play the main role as an effective aid in preventing heart infarctions, liver and gall bladder disease, arteriosclerosis and tumorous growths. Still more simply expressed, they act as an aid towards a healthy way of life.

When in 1954, I lectured in Karlsruhe on these scientific findings, a Japanese gentleman, the head of the Tokyo doctors' group, asked for permission to speak and he said: "She is right. In Japan we say that the decline of the Western standard of work is an internal, strangling thing, brought about by what they eat." We could stay healthy by using the human privilege of being able to practice judgement in the choice of what we eat, in order to maintain our physical mental and spiritual health.

The food we eat is not mankind's only determining factor. The body, soul and spirit all have their functions and roles to play, their areas of influence. But the harm done by eating the wrong kind of food fats has repercussions in all realms of life, including healthy mental and spiritual functioning. In our world, however, the choice of healthy food is one of the elementary aspects of our lives which we should organize as such. It is far more important than many people in the Western world are willing to admit. It is not those who acknowledge this fact who are materialistic in their way of thinking, but those who are not willing to forego something, in order to achieve a far greater goal.

SOLAR ENERGY AGAINST CANCER

Biological Prevention and Healing of Cancer

The Bio-chemical Reactions of Unsaturated Fats

Electron Biology and Resonance Absorption of Solar Electrons

A lecture held on 17 June 1966, during the Second Bio-technical Week in Neviges.

May I express my thanks to Mr. Kokaly for the warm invitation to speak in your circle. I am particularly pleased that the sun is smiling so kindly down upon us today because, in the next few months, you will see the sun suddenly becoming a very 'hot item'. While my latest book: *"Kosmische Kräfte gegen Krebs"* (Cosmic Powers against Cancer) was being printed, there had already appeared small hints in various journals, of how important solar electrons are. The book, *"Sonnenenergie und der Mensch als Antenne"* (Solar Energy and Man as Antenna), which has been announced, offers, as can be judged by the title, a theme for much discussion and pleasant anticipation. It appears that people do react very positively to the sun, despite the fact that many doctors today advise patients to avoid it.

Is the sun suddenly no longer of benefit to us? I do not believe that man's interventions in cosmic relationships, biological processes and our biological-dynamic balance, have gone so far as to negate the positive influence of the sun. That has to lie in the receiving antenna. It is quite possible that the human antenna for sunbeams is no longer functioning. Let us contemplate these relationships for a while, today.

This morning, someone who had said in the discussion yesterday—we are here, after all, thinking individuals: "Until now, we have always only thought over what has already happened; it is important that we perhaps even start to pre-think what could happen, but above all, we should think!"

That is my opinion absolutely, because, as we look at ourselves and how far man has followed various commercial interests and carelessly interfered with his biological-dynamic balance— thereby cutting off his own life-nerve—it is then clear that man as such, has seldom in his history, been so challenged as he is today to think hard about what being a human means, and to use once more, as Homo Sapiens, his intelligence. While I was giving my lecture: *"Über die Störungen des biologisch-dynamischen Gleichgewichtes in der Natur, und was der Mensch tun kann, um dieses biologisch-dynamische Gleichgewicht wieder herzustellen"* (Concerning disturbances of the biological-dynamic balance, and what man can do to restore it), at your colleagues' last conference in Gengenbach, we then had Professor Wellenstein, Principal of Freiburg *Forstzoologischen Institut* (Forestry and Zoological Institute), with us. I have reported from his publications how we are severing our own life-nerve by using pesticides.

In my lecture today, "What can we do to prevent and cure cancer?", I have formulated the theme rather more strictly. With special reference to the disease of cancer, I would like you to consider from the outset, that every interference or intervention which disturbs man's biological-dynamic balance, his place in the cosmic scheme of things, in the dipolar field current of electro-magnetic powers which surrounds the world and its creatures and which govern the entire cosmos; that every interference with these far reaching relationships, promotes the disease of cancer: The small section which I wish to address today, "What can we do to prevent and cure cancer?" is really only a partial extract from the enormous whole; as cancer itself is only the most advanced developmental stage of this interference in healthy life.

So, when I say that I wish to speak of the healthy vital functions in man, within the cosmic scheme of matter, radiant energy and electro-magnetic fields, it then means the same as when I speak of the prevention of curing of cancer. It is as simple as that.

At the end of my little book, which has just come out, *"Kosmische Kräfte gegen Krebs"* (Cosmic Powers against Cancer), in which I deal with sunbeams especially kindly, I wrote: "In the future, cancer research—I am firmly convinced of this—will become a very simple matter, clearly and easily understood by everyone." The best and greatest thinkers, not least those from the realm of physics, have, like Max Born, emphasized: "There is no better method of determining the basic laws which govern us than by the ideal one of using the greatest possible simplicity." Max Planck, quantum physics great founder, said: "When someone thinks he has discovered something new but he cannot as a scientist so express it that everybody understands, then he hasn't discovered anything new at all." This is why I am so firmly convinced that those who really have something to say which concerns all of us, can so express it that everybody understands.

What is all this connected with? Professor Wellenstain, the Principal of Freiburg's Forestry Zoological Institute has said of himself, as published, that he is one of those people, perhaps even the main one in Germany, who has done more damage than any other, but at last, his Saul has seen the light and become Paul. He is horrified at how, when he says he can no longer keep silent about the number of vintners who spray with calcium arsenic and are suffering from cancer, or about how many forestry workers who use DDT or Trinitrophenolin as insecticides are likewise suffering from cancer, the responsible authorities keep repeating that it is not his concern. It is the task of the Forestry Authorities. Professor Wellenstein says he can no longer close his eyes while the forestry and farm workers in this country are given such deadly poisons—insecticides—to handle, when alone in Stuttgart (according to Eichhalz) there are 120 official deaths from this cause. In larger districts for example, in a particular one in Dornap in the Rhineland—where insecticide was used over an area of 25 hectares, all the nesting birds died and 30% of the larger birds moved to other areas or died too. Deer, game, foxes and the like, were all found dead. It was observed—note the connection—that these dead creatures had all stored in their body fat, a wide range of insecticides, whether trinitrophenolin, Cresolin, DDT, calcium compounds or other insecticides used against oak pests. They are also stored in our own body fat, and have been proved to interfere with the living body's ability to assimilate that fat.

A similarly harmful aspect is found in the fattening-up of stock. While Senior Consultant Expert for fats, I investigated the high temperature treatment of fish oils, for the purpose of making them keep longer, and killing their fishy taste. I came to the conclusion that these oils then do great harm to the entire internal glandular system, as well as to the liver and other organs and are therefore not suitable for human consumption.

This was my official verdict prepared, by the way, for the *Ernahrung-Ministerium* (Ministry of Foods) in 1951. In 1955 I received the answer: Banning these heat-treated fish oils is "being considered" as tests in other institutes have now shown that these fish oils are very harmful to both human and animals, as they disrupt the functioning of the glands and poison the liver, which rapidly leads to death (Today, these fats are still commercially available!).

When, after 50,000 tonnes of the above mentioned fats had been bought by the margarine industry and, in this situation, selling these fats became difficult, a cattle-feed firm was quickly called into being for the production of so-called 'high-energy' for fattening pigs. This feed cake contained large amounts of barley edible, heat-treated fish oils, together with bone meal. Later, 50% of the pigs fatted on the feed cake turned blue and died on the way to the slaughterhouse. Furthermore, I also know of farmers whose young cattle have been harmed or have even died through such 'high-energy feed cake'. Through my consultation work with cancer patients I have met many trades people such as butchers, poultry breeders and bakers and, because of the situations which brought them such distress, I have learnt even more of these facts. I know that many butchers who are real experts at their job, are themselves aware of the fact that the meat, in the condition it often reaches them nowadays, is no longer fit for human consumption. If we, then, by the indirect method of such unbiologically fattened livestock, our-

selves ingest these harmful substances which act as inhibitors on
the fats metabolism, it should come as no surprise that this
behaviour, based on a greedy addiction to "getting rich quick"
boomerangs on us and we, ourselves, destroy our life-nerve.

How is it that these fats in particular are supposed to be so very
harmful and wreak such havoc? In official cancer research,
a book published by the well-known Nobel Prize winner,
H.V. Euler of Stockholm, states that if we were to reduce the
many and varied forms of cancer to a common denominator, we
would then have to say that the living body lacks the ability to
assimilate fat. We have to discover the factors which enable the
living body to once more cope with and integrate fat into the
vital functions. The 'lipotropic substances' became a very topical
theme in medicine. At that time I was the Central Government's
Senior Expert* both for fats and pharmaceutical drugs. As I had,
in 1950, already developed new and sensitive methods, the very
first of their kind, for detecting fats—before that time it had been
impossible to differentiate between saturated and unsaturated
fats—I was spurred on to investigate which fats had undergone
testing in 'lipotropic substance' studies. 'Lipos' means fat.
'Lipotropic substances' are intended to enable the living body to
mobilize and activate fat again, to make it more soluble. I
determined that the nature of the fats themselves had been
completely ignored. It was mostly fat bacon which had been
given in the animal tests. My findings that: Fat is not simply fat,
led to focusing on unsaturated fats. I succeeded in proving, on
a small piece of paper, that those protein substances which had
been used to reactivate the living body's ability to assimilate fat,
possess a 'haptophar', an adhesive capacity for fats, a sulphurous

---

*In the year 1952, now forty years ago.

connection. Electrically, this has a positive charge and belongs to heavy matter. With this haptophar, the protein substances also have a corresponding haptophar in the fats. It is called the electron-rich, double-unsaturated connection of fatty acids, e.g. linoleic acids. I put a drop of linoleic acid—15 years ago it had become modern—a minute spot of 1/1000TH of a milligram, onto a piece of paper, and then allowed the various 'lipotropic substances' to rise in the paper. I compared the reactions of these substances, e.g. the linoleic acids in the seed oils, and the reactions in pork fat. The results showed that in the seed oils there is always present, in varying amounts, the haptophor for the lipotropic substances. The lipotropic protein connections, e.g. Cystein, as they are found in Quark, cottage cheese or nuts are able to make water-soluble the biological highly unsaturated vegetable oils in seed oils. And that is what matters. When you mix together Quark or cottage cheese and linseed oil in your blender the fat becomes water-soluble. If you prepare the same mixture with fat from a pig which has been fattened on totally wrong feed, the fat no longer combines with the protein. It separates. You can see this as an illustration of chemistry, in vitro, of what takes place in all the organism's vital functions, and in its capillary activity, or in the rising of a plant's sap, that fat and protein *do* combine. That fat does become water soluble—but only when protein is in combination with highly unsaturated fats. Those highly, unsaturated fats are, essentially, rich in electrons. This is clearly measurable, physically.

These electrons enable fats to be surface-active at capillary level. Capillary activity is, quite simply, enormously important. In the centre of the earth as in the centre of the atom, there is much heavy matter. In the sky and in the outer envelope of the atom there are active, orbiting electrons. When we eat, electron-rich foods tend to move towards the surface which is of extreme importance for all the vital functions for the secretion of mucous, for the capillary activity of both the blood, the lymph fluid, and also for excretion through the bladder and intestines.

I often take very sick cancer patients away from hospital where they are said to have only a few days left to live, or perhaps only a few hours. This is mostly accompanied by very good results. The very first thing which these patients and their families tell me is that, in the hospital, it was said that they could no longer urinate or produce bowel movements. They suffered from dry coughing without being able to bring up any mucous. Everything was blocked. It greatly encourages them when suddenly, in all these symptoms, the surface-active fats with their wealth of electrons, start reactivating the vital functions and the patient immediately begins to feel better. It is very interesting to ask how this sudden change is possible. It has to do with the reaction patterns, with the character of electrons. I will return to these electrons later. In the last two years I have come to be very fond of them. A friend of my work in Paris wrote to me how wonderful it is that you have discovered the original birthplace of the electrons in seed oils to be the sun. That's how these connections are made!

We must ingest these electrons in our food for the vital functions—actively striving towards the surface—to once more function as they should, thus making us feel lighter in ourselves. Some people think that they work too much. I no longer accept this from most people because, when I ask how long they actually work, it is most certainly neither too long nor too hard.

They are of the opinion that they work too much because they feel heavy and tired and keep wanting to lie down. When patients of mine—even very sick cancer patients—have, after consultation with me, followed my "Oil-Protein Food Plan" for as little as only one or two days, they, these people who have never read a single book of mine, and without knowing my theory, come and quite spontaneously report: "I suddenly feel so light in myself, no longer so heavy."

How is it that electron-rich nutrition makes one so light? Why do we need it so urgently? Electrons consist of light matter. They take away weight from heavy matter, which drags us down. They lighten the pull of the earth and makes us suitable for the heights.

Electrons have a great affinity for oxygen—they love it. That is why, in us, they thrust for the surface. They attract oxygen and stimulate our breathing—our entire being.

Electrons also quickly lead to decay in many foods, particularly when the natural preserving agent of the seed or fruit is damaged. The destruction of the light, activating, electron oxidation system is the reason behind preserving processes which treat food to make it keep longer, and change the nature of fats so that they can be stored for years.

Such food acts as if it were stone instead of bread. It loses the feature of being properly nourishing. Testing on animals and humans has shown that when these preserved fats, poor in electrons, are ingested, animals and humans eat six times their normal amount of food. According to this, as I often tell housewives, they could save ⅚ of their housekeeping money, if they chose the right kind of fat. That also applies, I assure you, to this house here. Man's hunger for meat is only proof that he has lost his instinct for the right kind of nutrition. I do not forbid patients meat, unless they are on the edge of the grave. I do not eat meat. Someone wrote to me from Switzerland that they know I do not eat meat myself, but allow patients to do so, and he asks why? I answered: Kind-heartedness. But it is better, especially nowadays, to live without meat. It is good to observe that people who eat the right kind of nutrition—electron-rich foods—tend to avoid heavy food, preserved food, and interestingly, these people gradually begin, of their own accord, to stop wanting meat, until they finally say that meatless nutrition does them more good.

What does such food consist of? The "Oil-Protein Food Plan" I have developed contains electron-rich oils, seed oils and beneficial protein which, in combination with the oils can activate the body's vital functions. Additionally, this food plan contains large amounts of fresh vegetables and fruit. Anything can be eaten which has not, through man's chemical intervention, been blemished or 'deactivated', leaving only the food's heavy matter. Why is this simple question of nutrition in connection with fats so important? In 1911 a Swedish researcher had already published: "Fats are the substances which govern all aspects of life". 'Protein' is unjustly called 'protein'. Proteos = I come first. A cell's living and dying are revealed in how fat and protein associate within it. This is also known form narcosis research. All the poisons which harm the living organism narcotically or toxically, including the benzopyrin in cigarettes, are stored in the fat and divide the fat and protein association process. The following animal tests are very conclusive: When animals breathe in relatively large amounts of cigarette smoke i.e. benzopyrin; it is proved that cancer does occur. In parallel tests, animals were concurrently fed with flax seed or flax seed oil. These animals did not become ill. By that, I'm not telling you to take up smoking heavily. With this example, I want to indicate how important it is to recognize not only where harmful and poisonous substances affect us, but also how these effects can be overcome. The latter aspect seems to me more important than simply busying ourselves with the "toxic situation overall". At present we can neither hinder radioactive fallout, nor make the margarine industry, with a yearly turnover of *21 Milliarde Deutschmarks,* change its fats production methods. These people say: "If we were to dismantle our machinery, it would cost us millions. Who's to give us money to build something new?" What we can do personally, for ourselves, our surroundings and environment, for our friends, is find a method which helps to overcome most of this harm and damage. Today, just because it can mostly compensate for the above mentioned harmful influences, electron-rich nutrition is more important than ever before. I have explained in detail the reasons for these links between the detoxification functioning of the body, and the water solubility of electron-rich fats in the booklet *"Kosmische Kräfte gegen Krebs"* (Cosmic Powers against Cancer). I already emphasized, 15 years ago*, the limiting effect on breathing when there is a lack of electron-rich fats, in which case, all glandular functions are reduced, and the secretion of mucous in the body's upper and lower cavities, including that of sexual functioning, are choked. This is also of importance in stock-breeding. It is worthwhile considering whether it is really

*In the year 1951 & 1952

profitable for farmers to give harmful feed cake in an attempt to make quick money, but which leads to infertile stock. Some listeners here today will be interested to hear that people, just like you and me, as two individuals, couples too, can solve the problem of marital infertility by changing the foods they eat. A farming family in Buchau on Lake Federsee had been wanting a son for over ten years. Would you believe that ten months after they consulted me on food and nutrition, I received a photograph of their first son and heir!

These related aspects, standing as they do, at the very start of life, are affected by the cardinal question of the interplay between fat and protein. Here is one example of how important this inter-reaction is, this central function of the association of fat— biological unsaturated fat—and protein, in connection with the division of the cell and every other vital function: The male spermatozoa contain over a thousand times more sulphurated protein than any other cell. The female ova are, due to their lecithin content level, rich in highly unsaturated fats and there- fore determine the female sex. Warburg has described how, at the instant of fertilization, respiration increases a thousandfold and how the synthesis within the ovum, like all structure formations of its kind, is in this sense, as we see now, dependent on the interplay between fat and protein.

As a result of my work's incorporating modern quantum physics, which I have done during the last few years, evolutionary, and enormously important and beautiful perspectives have been revealed. The physicist Dessauer has written that new work in Munster (done, I may add, by me personally), indicated that the architecture of the living molecules between heavy matter, between protons and the light matter in the electron, can attain a certain peak, according to the developmental level. The more developed plants or animals are—up to and including human beings—the richer the substance is in its level of electrons. No animal is so badly harmed as is man, when electrons are withheld. Conversely, it has been proved in the Zoological Institute that, through the chain of concentration that exists, the dangerous effects of insecticide increase with the individual's level of higher development.

When a thrush eats a worm with a high amount of stored DDT in its body, the thrush will die, although the worm was able to survive with such a concentration. When a bird of prey eats a mouse which has stored a lot of DDT in its organism, but which still survived, the bird will die. In forests where insecticides have been used, there is less poison in the water that in its plankton, there is more poison in fish than in plankton—and when a bird eats such a fish, it too, will die. The more highly evolved a creature is, the more sensitive the animal, the organism, the living body and thus the human is, to the withholding of electrons or to the harmful effects of chemicals which disturb the exchange of electrons. The system of unsaturated fats in association with protein is in fact a dipolar system in its nature, where a continual exchange or movement of electrons is able to take place.

And now for something really beautiful! In this bipolar system between highly unsaturated electron-rich fats and protein, the presence of sunlight activates a depositing and storage of solar electrons. Dessauer writes: If it were possible to increase the concentration of solar electrons tenfold in this biological elec-

tron-rich molecule, then man would be able to live 10,000 years.
It is, in any case, true that all so-called mutations which damage
the genetic factor, which have a negative effect on man and
which cause disease, indicate a withholding of electron energy,
while the concentration of solar electrons—stored in the bipolar
molecule between unsaturated electron-rich fats and protein—
can greatly increase the depository effect of strength and power
as well as the subjective health condition. A wealth of electrons
means an increase in happy well-being. This result of a positive
sum of strength and power was also recognized by Aristotle.

I keep observing that the subjective condition of those patients
I have treated with massive doses of flax seed oil—sometimes I
give an enema of 500CCs of an oil mixture right at the start
of therapy—that their subjective awareness of well-being,
increases immediately. The concentration of electrons means an
increase in man's feeling of happiness. Everything is relative, this
statement also, but it is still correct. We can include Einstein's
theory of relatively here. These interconnections, these facts, can
be readily proved.

Why is that? The living mass of mankind derives its *being*—as
does all life in nature—from the sun! This has been forgotten
until now in biology and medical science. The living body can
only take in and store solar electrons through resonance absorption. To absorb the electrons into the living body, we must
already have in the body's electron system either the same
wavelength or a multiplicity of wavelengths. Thus, the human
who eats refined foodstuffs or food which lacks electrons, not
only cuts off his oxygen enough to suffocate himself, he also cuts
himself off from the effects of the sun. When such people cover

their skin with a layer of paraffin as sun protection oil, and then lie in the sun, the burn damage is very great, because the electrons, which cannot be stored, and the electron-rich biological molecule, are missing. It has been proved that all the poisons which affect the action of enzymes, including paraffin and the benzopyrin in cigarettes, have an irritative effect on the entire system of electron absorption, storage and further conduction.

This storage depot of solar electrons, of energy—which in turn increases the earth's bio-energy levels—is brought about by the action of sunbeams in the biological living body—and can be drawn upon and passed on according to the situation and demand. In clarification of these inter-relationships, modern quantum theory in physics is of particular interest and highly illuminating. Earlier there were two opposing opinions concerning the nature of solar energy. Newton said it was composed of corpuscular beams and therefore material. The other interpretation, which we looked at yesterday, emphasized that it was a matter of electro-magnetic waves. This light theory comes from Huygens. Through the establishing of quantum theory by Max Planck and further development and work done by Einstein, and last but not least, through the work of the Frenchman, Louis de Broglie, the following is incontestably recognized in physical science today:

Solar energy electrons are both wave and matter! De Broglie writes that light is the fastest, purest, lightest and most beautiful form of matter we know, as well as the fastest and purest form of energy we are aware of. As the fastest emissary from star to star, sunlight electrons are always, whatever their condition, both wave and matter. The electron is a form of matter always surrounded by magnetism: according to the measurement methods used, it can be measured as either matter or wave. This borderline situation between energy and matter overturns all classical physics, and is extraordinarily interesting as well as of vital importance in respect of physiological, medical and biological problems—and will, I am very confident, give our biological goals and our biological thinking great support and a strong impetus. We can store the sun's energy and the living body is then in a position to summon, depending on the situation, energy from this storage depot of electrons. When these depots are empty, the person then feels irritable, tired, and his limbs become heavy. But we are able to replenish these storage depots by taking in electron-rich seed oils. These are set to receive solar energy.

You cannot, by selective seed cultivation, force the level of unsaturated fats in oil seed to rise, without also raising its level of protein. The harmony between fat and protein, between beneficial fats and beneficial protein is of immense significance for all the vital functions and life processed in this dipolar, biologically important system which controls everything in the life processes themselves. There, in the harmony between beneficial fat and beneficial protein, there is a continual easily influenced and directable biological molecule prepared to store the sun's radiant energy, to absorb and radiate it, according to the individual's need for energy, strength or power. Without any doubt, every function of the brain—and this has been scientifically proved—needs the very easy activation effect of threefold unsaturated fats. A Swede has proved that no brain function can take place at all without threefold unsaturated fats. The same applies to nerve functions, and for regeneration within the muscle after strenuous muscle activity, in the so-called oxidative recovery phase during sleep. This process requires the highly unsaturated, particularly electron-rich fatty acids in flax seed oil. So, when I wish to help a very sick patient, I must first give the most optimal oil I have. Given the climate in which we live, my opinion is— Flax Seed Oil. All other oils can be given later. In every case they should be natural, untreated oils.

Why is the other kind of radiation which is given to cancer patients today, wrong? This, in the light of physics research, is of great interest. At an International Radiation Therapists' Congress—very up-to-date—which took place in Switzerland in 1956 and brought together scientists from all over the world, it was reported that the effects of electrons had been tested in a solution which contained dead, inert matter dissolved in water. The physicist Dessauer, DR.MED.H.C. and DR.THEOL.H.C. has written that the absorption of all radiation, sunbeams and the other cosmic rays which can be measured in the living body as well as in inert areas, differ in a fundamental way. Cosmic rays which contain protons, heavy matter or neutralized electrons are neither absorbed nor conducted by the living organism. The organism has a kind of filtering procedure for separating suitable and unsuitable radiation. When the body is exposed to too much unsuitable radiation, burn damage results. These unbiological, unsuitable kinds of ray, whether from synthetically manufac- tured radioactive substances such as radium or cobalt, or in the form of unfavourable cosmic rays, only serve to destroy the living body. All attempts, undertaken by doctors, to localize the currently employed form of radiation treatment to the tumour are bound to fail due to the basic nature of the physical occurrence taking place, which is that the wavelengths of these rays find no answering resonance in the wavelengths of our body's own biological electrons, which we are able to store in quantity from the sun. These 'black rays', according to Max Planck and the 'incompatible rays' according to Dessauer, are in no way to be used as therapy on any living organism. At the Cancer Research Congress in Moscow in 1962, this was also publicly emphasized by the President of the Scientific Committee. There were 150 German scientists present. We had a long discussion on this over Krimsekt in the German embassy, at the invitation of Ambassador Kroll. The defenders of the old, obsolete methods which use the wrong, injurious kind of rays are in fact aware of these new scientific findings and related comments made by the Head of the Scientific Committee at the major congress in Moscow: We have to see that radiation treatment and operations are regarded as obsolete in our era. We do not find these methods convincing and there is not one scientific argument which justifies the use of this kind of unbiological radiation. We have to turn to immuno-biological methods. It has been scientifically proven (by Professor H.V. Euler of Stockholm), that every operative measure undertaken in cancer cases only make the end more unbearable, causes faster formation of metastasis and can no longer be at all advocated. *[Spontaneous applause followed.]* Nowhere in the world, and I have been in Japan, China and India and have, I think, attended science congresses in most European countries, and maintain contact with many researchers in this field, and I say: "Nowhere in the civilized world, nor in those countries which are still underdeveloped, is the movement in the direction of biological treatment fought against or brought to trial in court, as often as it is in Germany. The instigators are a number of die-hard dictators in the medical institutes. Here we have a treatment; recognized as wrong, but which is still given dogged and obstinate credence. I can say this because numerous representatives of this school of thought have been to me seeking help with members of their families. They found that help but still continued to support such obsolete methods in the medical journals." *[Great applause.]*

I, too, have often been taken to court. The cases were decided in my favour. All these cases were brought by various General Medical Councils. In Freudenstadt it was requested that the case should be held in another town, as the entire District Court was too biased in my favour. In Rottweil, I was once more acquitted. After a second hearing by the District Attorney, instigated by a General Medical Council, the matter was passed on to the Main Court of Assizes in Stuttgart for their verdict. An acquittal was also the result there. When juristic methods had no success, attacks were made through the newspapers. I had my chance to answer them. Through these public attacks, more patients than ever before came to me for consultation and treatment. I am now in the fortunate position of being able to say to you that among general practitioners, most of them, the sensible ones—so many are so wide-awake these days—firmly declare for what I do. Only the institutions and their managers still defend the obsolete methods.

And what do I actually do? I give cancer patients simple, natural foods. That is all. I take sick people out of the hospital, when it is said there that they do not have more than an hour or two left to live, that the scientifically attested diagnosis is at hand and that the patient is completely moribund. In most cases I can help even these patients quickly and conclusively.

What can we learn from this? Firstly, when the simple restoration of our fundamental nutrition—oil, protein, fruit and vegetables—is so very important, it is up to us to see that they are on offer in unadulterated form. Secondly, when maltreatment by chemical means, preservation methods, antibiotics or refining processes make certain foods such a source of danger, we should take great care to avoid them.

A point of cardinal importance in the maintenance of healthy vital functions has been found in research into the metabolism of fats and its connection with the taking in of oxygen and the absorption, storing and activating of the sun's radiant energy.

From the cardinal viewpoint of the body's vital functions, one suddenly becomes aware of just how dangerous is so much of our interference in the biological-dynamic balance of Nature. It doesn't matter whether it has to do with methods of insecticide usage, or with the antibiotics given in the fattening of stock, or with methods of preservation which utilize respiratory poisons or with the destruction of our wealth of electrons. These methods all turn fats into breathing inhibitors, when they should be breathing activators. The fats then become storage depots for these poisons, becoming in that action of storing, a disturbance factor which acts in opposition to the storage process of life-giving solar electrons. This means man is brought low, degraded and hindered in his positive evolution.

If we want to survive, as a species of man and as an Occidental race, and to live in contented happiness, it is then time for us to reconsider the biological fundamentals of our being. A Chinese principle in research and philosophy is: "The secret of all wisdom is science in harmony with the ways of nature." This spirit of learning urgently needs a revival in Western science, in the opinion of Radhakrishnan. It is high time for us to finally start thinking, and acting, like responsible beings.

[Long-lasting applause.]

THE FATS SYNDROME & PHOTONS AS SOLAR ENERGY

A lecture held on 6 April 1972 at the 8th "Vie et Action"
(Life and Action) Congress in Tours, 5–9 April 1972.

Sunbeams, photons, electrons. . .What are they?

Sun rays reach the earth as an inexhaustible source of energy.
The sources of power in mineral oil, coal, green plant-foods
and fruits are based on the energy supplied by the sun's radiation. In 1960, a Japanese physicist published the following:
"God was indeed the Creator when he said: Let there be
light!" Your fellow countryman, the holder of the Nobel
Prize and a highly deserving physicist, Louis de Broglie,
writes: "if we let our imagination run free, we can easily
picture that at the beginning of time, the morning after the
divine 'Fiat Lux', light, unique in the world, with gradual but
continual concentration, created the material universe we
today acknowledge around us. And perhaps one day, when
the time has come to pass, the universe will regain its original
purity and once more dissolve back into light."

Light is the fastest emissary from star to star. There is nothing
faster than light. Light rushes along with time. It lives eternally.
Physicists emphasize too, that the photon, the quantum, the
tiniest part of a sunbeam, is eternal. Life is unimaginable without

the photon. It is in continual movement. Stopping this is not possible. The photon is filled with colour. It can, when present in large numbers, change colour and frequency. The photon is recognized as the purest form of energy; the purest wave and, in continual movement, it can combine in resonance with a second photon to form a 'short-life' particle. This particle—known as an "O" particle—can once more break itself down into two photons, without mass, as pure wave in movement. This forms the basis for the wonderful interplay between light and matter. It is not possible to pinpoint the location of a photon. The theory of relatively rests on this foundation.

The photon, the smallest quantum of sunlight, gave rise to Max Planck's and Einstein's formation of the quantum theory which is of such significance today.

This photon, so active, so dynamic, so powerful, can be captured by suitable electrons. What does this mean?

Electrons are already a constituent of matter, even though they are also in continual movements. They vibrate continually on their own wavelength. They have their own frequency, like radio receivers which are set at a certain wavelength. The electron orbits in matter around a nucleus. The heavy matter in the nucleus is charged with positive electricity. In contrast to this, the electron carries a negative charge. The two; the positively charged nucleus and the negatively charged electron attract each other by means of their electrical opposition. But the electron, always in motion, never approaches the nucleus close enough to be drawn out of its own orbit. It maintains a certain freedom of movement within its prescribed orbit. Electrons love photons. They attract the photons by means of their magnetic

fields. When an electrical charge moves, this always produces a
magnetic field. The moving photons also possess magnetic fields.
Both fields; the magnetic field of the electron and that of the
photon, attract each other, when the wavelengths are in tune.
The wavelength of the photon—which it can change—must so
fit into the circular wave of the electron that the orbit is always
completely filled by the wave. This feature is, when we consider
its physical expression, its biological and even its philosophical
consequences, extremely interesting. This basic physical law,
according to Niels Bohr, means that no matter can change its
condition without absorption or radiation of electro-magnetic
waves—and the photon is an electro-magnetic wave—without
the wavelength corresponding to that of absorption or radiation
frequency—to its own radiation frequency.

Matter always has its own vibration, and so, of course, does the
living body. The absorption of energy must correspond to one's
own wavelength.

Sunbeams are very much in accord with humans. It is no
coincidence that we love the sun. The resonance in our
biological substance is so strongly set to absorb the sun's energy
that those physicists, the quantum biologists, who concern
themselves with this scientifically complex field, say that: There
is nothing else on earth with a higher concentration of photons
of the sun's energy than man. This concentration of the sun's
energy—very much an iso-energetic point for humans, with
their eminently suitable wavelengths—is improved when we eat
food which has electrons which in turn attract the electro-
magnetic waves of sunbeams, of photons. A high amount of
these electrons which are on the wavelength of the sun's energy,

are to be found, for example, in seed oils. Scientifically, these oils are even known as electron-rich essential highly unsaturated fats. But, when people began to treat fats to make them keep longer, no-one stopped to consider the consequences of this for the existence and higher development of the human race. These vitally important amounts of electrons, with their continual movement and wonderful reaction to light were destroyed.

It is interesting that the science of physics has been marred by the term: "anti-Mensch" (anti-human). We will return to that later. It is the human being, with his gradual concentration of electrons, ever striving towards the future, who conceals within himself the greatest potential for the sun's energy on earth. The mirror image of this, a human lacking electrons, lacking photons, indicative of the past, perfectly illustrates, physically, the anti-Mensch. There will be more details of that too. At the moment we are considering the wonderful relationship between man and the sun's energy.

The sun's energy and man as an antenna: almost everyone knows what an antenna is. The marvellous findings of Maxwell the physicist, concerning electro-magnetic waves in technology today are well-researched and of practical use. Famous examples are telegraphy, radio, television and various other applications of high-frequency technology in the manufacturing of electro-magnets, the atom bomb and research into nuclear power as a source of energy.

These technical developments and their relation sciences are based on scientific findings which go back to the nineteenth century. As important milestone came in 1888 when Hertz published his: *"Ober Strahlen elektrischer Kraft"* (Concerning Rays and Electrical Power). Using mathematical equations, Maxwell was able to prove that connected electrical and magnetic fields, when in movement, radiate electro-magnetic rays. These elementary laws also govern processes in nature.

An electrical charge in movement produces a magnetic field. Electrically conductive matter which is moved within a magnet's field will produce current. When connected electrical and magnetic fields are separated, the ensuing phenomenon gives rise to radiation of electro-magnetic waves. These fundamental, elementary laws can also be applied to biological processes.

When the sun shines on the leafy canopy of a tree and is absorbed through photosynthesis, this causes movement in the electrical charge of the electrons. A magnetic field is also brought about when the water in trees rises. When we, with our wealth of electrons and conductive living substance, move through the electro-magnetic field of a forest, then a charging with solar electrons takes place in us. When our blood circulates, there is a movement of the electrical charge in the magnetic fields (for example, on the surface lipids of red blood corpuscles), which then causes much induction and re-induction of energy.

With each heartbeat, a dose of the body's own electron-rich, highly unsaturated fats from the lymph system, together with lymph fluid, goes into the blood vessels and thereby into the heart. This constantly stimulates and strengthens anew the electro-motoric functioning of the heart. Even, the movement of the bloodstream is connected with radiation of electro-magnetic waves—in accordance with the fundamental law of nature which governs electro-magnetic waves. This transmitter within humans is always in action.

The tubular arrangement of our nerves with their various layers and nodes, their different potential, neurons and dendrites, all immediately convey a picture of how strong an electric current in a magnetic field has to be to lead to the radiation of electro-magnetic waves here, too. When I begin thinking something positive about a person, this is also related to a radiation of electro-magnetic waves. Reception of this is also dependent upon the wavelength the receiver is on—and there are also amplifiers as well as transmitters of interference. All this brings in a large amount of data, which is commonly known under different names, such as telepathy, hypnosis, thought transference and many others.

Among the northern races, it is recognized that isolated native inhabitants use trees to help them to send their thoughts over distances, to inform a husband for example, who has gone to town, that he has to bring some salt. Bismarck wrote about how, during times of affliction, he would wrap his arms around a tree, lean his forehead on it and find relief and relaxation. In both cases, it is a matter of electro-magnetic waves in accordance with the mathematical equation of Maxwell.

But our theme is not really in this mental sphere, but has to do with measurable waves. Let us concentrate on the actual fats syndrome in its effects on the brain and nerve functions, the organs of the senses, the secretion of mucous, the functioning of the stomach and intestinal tract, liver, gall bladder and kidneys, the lymph and blood vessels, the skin, respiration, the immunity system, the fertilization processes and sexuality, semi-narcosis or vital power and also in connection with growth processes—experiments have proved conclusively that all of these systems and processes of the human being are very much connected with electron-rich highly unsaturated fats as receivers, amplifiers and transmitters of electro-magnetic waves, and as overseer of the vital functions. It is here that the sun's energy is an active influence, radiated by the powerful photons as purest movement or controlling the electron structure of the living body, either bound to matter, or as power and energy from electro-magnetic fields and waves.

The dynamic of the vital functions and their basis in energy from the sun is, today, a fact which has to be reckoned with. The dynamic effect of sunlight on the body's vital functions has already been observed by an alert professor of optical medicine, Prof. Holwich. Blindness is often followed by a deterioration in many functions of the internal organs and in the secretions of the liver, gall bladder and pancreas. The heart's action is also affected, as is the peristaltic motion of the intestines. When sight is regained, an unmistakable activation of these organic functions is observable. You could say that this was due to physiological factors, for instance, to better and more movement out of doors, or to psychological motives. But it is worth mentioning, in connection with this, admittedly secondary observation, the following fact: A young doctor, a hunter, observed that in some years, prize-winning deer antlers were much larger than usual, but in others, even the very best of the prize-winning antlers were far below average in size. The physical development of young animals and ducks in the wild matched in degrees the size of the deer antlers, or lack of it. Investigation showed that the years in which the deer grew large antlers and in which there was good physical development of young animals in the wild, were clearly years with high rates of sunshine hours.

That the sun, as the element of life, also affects the dynamics of the vital functions is apparently self-evident. Everyone seems to agree. But then why are there so many people who say: I can't tolerate the sun? The answer is on its way. Firstly, a few words on the topic of: The dynamics of the vital functions and their base in the sun's energy: When I have treated patients and they then lie in the sun, these sick people notice that they begin to feel very much better; rejuvenated. In contrast to this you often hear about people on sunny beaches having heart failure. It is not unusual for heart infarction to occur. Both conditions are observable. For some people nowadays, the sun's energy is an overly strenuous matter, while for others the dynamics of the sun's energy have an invigorating effect on all the vital functions. The stimulating effect of the sun on the secretions of the liver, gall bladder, pancreas, bladder and salivary glands is easily felt. These organs only dry out under the sun's rays when the substances which stimulate secretion in the body are lacking. With all these observations it is of decided importance, whether the surface-active electron-rich highly unsaturated fats are present as a resonance system for the sun's energy, or whether they are not. Doctors tell cancer patients that they should avoid the sun; that they can't tolerate it. That is correct. The moment, however, that these patients—cancer patients as well—have been following my oil-protein nutritional advice for two or three days, which means that they have been getting sufficient amounts of the essential fats, they can then tolerate the sun very well. Indeed they emphasize how fine they suddenly feel in the sun—how much their vitality and vigour is stirred and stimulated. And that brings us to the decisive point of this lecture.

The electrons in our food serve as the resonance system for the sun's energy. The electrons in our food are truly the element of life. Their electro-magnetic field attracts the photons in sunlight. These photons with their own vitality and continual movement, without which the physicist cannot imagine life existing, these photons, which are in resonance with the electrons in seed oils, are focused on the same wavelength as the sun's energy, serve the life element. This interplay of solar energy photons and the electrons in seed oils, which are focused to the very last quantum on solar energy photons, governs all the vital functions. Fats are the dominant factor for all the vital functions, according to Ivar Bang, a venerable physiologist from the year 1911.

Today we can add to that: Yes, fats which are in accord with solar energy photons, electron-rich fats in resonance with the solar energy wavelength, do indeed govern the entire vital function systems. In my work it is fascinating to see that my utilization of the relativity theory's related quantum physics illustrates how each and every individual vital function's dynamics are dominated by this exchange of electrons.

The electrons of highly unsaturated fats from seed oils, which lie on the same wavelength as sunlight, are capable of drawing solar energy and storing it, then, upon demand, of activating it as the purest energy of the dematerialized clouds of electrons, and making it available for the vital functions. All the vital functions are closely connected with membrane functions. The exchange of electrons, the distribution of energy in the whole organism is dependent on these membrane functions—in the nerve pathways, the brain, in every organ, the liver, gall bladder and pancreas in the stomach's mucous membrane and in the kidneys and intestinal tract: The controlling functions of these membranes with their electro-motoric power, is felt everywhere. This is also true for the respiratory functions, and in oxygen absorption and utilization. It also applies to cell division—to all normal growth processes. It is true for the catabolism of substance in the elimination processes taking place by way of the kidneys, intestinal tract and also for the growth of hair and nails, as well as for the development of young life in the womb. This brings us to a rather beautiful, and extremely important aspect of development.

According to the computable findings of those modern physicists, the quantum biologists, there is no entity in nature, in life, which has a higher concentration of solar electrons than *man*. It then follows that *man* has a true rapport with sunlight. Physicists today are recognizing more and more that: "Let there be light!" at the outset of Creation is becoming, physically, ever clearer to our minds, as the *truth*.

According to Einstein, the earth's gravity can be partially neutralized by means of electrons which conform to the sun's energy. The sun, and the sun-attuned electrons in the food we eat, bring us to a higher level of energy, and to a higher level of development as a human being.

"Anti-Mensch" and radiation damage are aspects which, through my research's utilization of modern quantum physics, have been shown to be actively promoted by means of radiation treatment for cancer patients. Physicists interpret from mathematical formulae that man, with his wealth of electrons, is directed forward in time. As we have heard, the photon speeds along in time; it has, so to speak, eternal life. Correctly considered in the abstract, proper mathematical formulae for important physical correlations can be so altered—still maintaining their mathematical correctness—that time is seen not to be directed forwards at all, but backwards. All one has to do is to give a negative sign to the time quotient in this mathematical formula and that is, of course, possible with standard mathematical rules. There is then for many physical 'particles' a corresponding 'anti-particle'. This was sought and found in the world of elementary particles, in the elementary particle's 'zoo', as it is known to physicists. By means of these mathematical formulae, applied to Physics, and by reversing time, the mirror image of human beings is delineated—the "anti-Mensch". While physically speaking, man represents the highest lever of order, that is directed against entropy, the anti-Mensch, according to physical formulae, lacks electrons. As I see it, the "anti-Mensch" is, physically and mathematically directed into the past. The "anti-Mensch" possesses few solar energy photons—the lowest level of order, physically speaking.

The process by which x-rays, gamma rays, atom bombs or cobalt rays are set in motion is also equally directed toward the development of the "anti-Mensch". The electronic structure of the vital functions is destroyed by such rays. According to Feynman's "World Line Diagram" and modern physics' theory of relativity, time and space have been given a relationship in a formula. The "anti-Mensch" is directed into the past. Mankind's internal structure with its interplay between solar energy photons and large number of electrons, with its concentration of photons in life's activities and in the dynamics of the vital functions, is directed into the future. This future-directed striving can develop powerful vigour. The "anti-Mensch", lacking electrons and directed into the past—his thought processes, too—is paralysed in his vital functions, lacks power and strength because the element of life, the sun-attuned electrons, are missing.

It is then fascinating to investigate our food, with regard to this feature. Fats and oils treated to make them keep longer have had their electronic structure destroyed—otherwise oxygen would be absorbed—and this has a very detrimental effect on human beings who with their wealth of electrons, live towards the future. This negative aspect concerning the development of the "Anti-Mensch" is in accordance with Feynman's "World Line Diagram". I emphasize that it means the fats and oils which have had their electron structure destroyed serve, within time and space, to promote the development of the "anti-Mensch".

Such fats, which interfere with the electron-exchange taking place in the living body because they act like insulating tar on the conductivity of electrons, also deaden the vital functions at the very commencement of their effectiveness—for example, in the organs and active growth centres as well as for the body in general. Tars were among the first substances to be recognized as carcinogens. What is cancer? Every single event in the "world of those elementary particles", which promotes the development of the "anti-Mensch" also promotes cancer. A high amount of heavy particles from the world of elementary particles in our food, the food we eat which is lacking a wealth of

electrons, promotes development of the "anti-Mensch". It
increases the occurrence of cancer. Solidified fats and oils for
example, belongs to this group. They lack electrons, acting as
insulating tar regarding the transport and concentration of
electrons in the living body.

Electron-rich food, electron-rich highly unsaturated oils,
natural flavourings from herbs and spices, from fruits which are
rich in aroma and natural dyes from the colour of the sunlight's
photons—these all help the absorption, storage and utilization of
the sun's energy.

How can you once more reach the peak of your development?

Freeing yourself from the influences and effects of radiation and
from environmental factors which promote development into
the "anti-Mensch", seem important. These goals, set by the
individual who chooses, or by the state and food industries with
their organisation and planning, should be to see that the food
we eat consists of electron-rich nutrition. An electron-rich food
intake which supplies us with the resonance system for the sun's
energy, must once more achieve priority. Such food, as the life
element, promotes our sun-attuned energy. This in turn promotes our development, in space and time, into the future. The
entire self can then grow and continue to develop further until,
in accordance with the laws of nature which govern light and
life, the highest level of our being is achieved.

Your fellow countryman, Prince Louis de Broglie, published
"Die Wellennatur der Materie" (The Nature of Waves in Matter).
His work brought him the Nobel Prize. The wave aspect of
living matter is all the clearer, the more one considers the nature

of solar energy photons. Sunbeams, in accordance with the natural laws of light and life, affect and govern the wave and radiant energy capabilities of our living human selves, today, into the future and forever.

Flax Oil recommended by
Dr. Johanna Budwig
is available at

BARLEAN'S ORGANIC OILS
4936 Lake Terrell Road
Ferndale, Washington
United States   98248

Toll Free Number 1-800-445-3529

Discover how to utilize the healing powers of Flax Oil in over 500 deliciously possible recipes in…

Dr. Johanna Budwig's

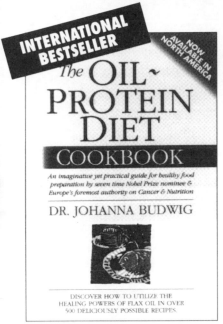

OIL-PROTEIN DIET Cookbook

An imaginative yet practical guide for healthy food preparation by seven-time Nobel Prize nominee and Europe's foremost authority on Cancer & Nutrition.

*T*his brilliant scientific mind has put together a guide for the use of oils in daily meal preparation…

- *Discover over 500 delicious meal possibilities using the healing powers of Flax Oil.*
- *Learn about 'good' fats and 'bad' fats and the proper use of fats in daily cooking.*
- *Select nutritionally important foods.*

- *Recipes for the young and young-at-heart—the convalescing, as as well as the athlete.*
- *Create exciting and nutritious desserts for the whole family.*
- *Find out how Quark and Flax Seed Oil offers optimum strength and energy.*

www.applepublishing.com